The potential for ground source heat pumps in domestic houses in the UK

Riyaad Moreea

Cranmore Publications

A catalogue record for this book is available from the British Library

ISBN: 978-1-907962-32-5

Published by Cranmore Publications

Reading, England

www.cranmorepublications.co.uk

This book is dedicated to my amazing mum who has always been there for me. I would also like to thank my dad, sisters, step-mum, cousins and friends who are all an important part of my life.

Table of Contents

Chapter 3 – Method

Chapter 4 – Data Section

Chapter 5 – Discussion

Chapter 6 – Conclusion

References

Preface

Ground source heat pumps are a very successful renewable technology used in many countries, so why are they less popular in the UK? With the current focus on going green, and the environment being a 'hot topic' at the moment, it seems that ground source heat pumps have been overlooked in the UK. They clearly have the potential to greatly help the UK to reduce its environmental impact.

This book looks into existing case studies of heat pumps in the UK to determine whether they should be widely installed in domestic properties. Different factors will be considered such as financial viability, carbon dioxide emissions and running costs, to see how GSHP can benefit the environment as well as the occupants.

With the governments tough targets to reduce the UK's carbon emissions by 15% in the domestic sector, surely ground source heat pumps can help achieve these targets? This investigation will analyse the environmental impact of the systems to see whether they can realistically help meet the government's targets.

Another important issue at the moment is the fuel poverty line, with current heating bills increasing and the trend set to continue, people are struggling to pay their bills. I will consider whether ground source heat pumps can provide affordable heating, notably for vulnerable people.

Chapter 1

Introduction

1.1 Introduction

Geothermal (ground-source) heat pumps are one of the fastest growing applications of renewable energy in the world, with an annual increase of 10% in approximately 30 countries over the past 10 years

Curtis (2007)

There is a lot of interest in this technology because it not only uses the heat found in the ground, which is a renewable source of energy available in most countries, but it also plays an important role in reducing primary energy consumption and carbon reduction. This is important because there are new government legislations starting to be introduced to reduce emissions such as the "Carbon Reduction Commitment."

The technology is well established in North America and parts of Europe, but fewer installations exist in the UK.

1.2 The history of ground source heat pumps

In 1912 Heinrich Zolly, from Switzerland, was the first person to patent the idea of using the ground as a heat source (Worth 1955). There was not a lot of interest in the idea at that time because the efficiency of pumps was not good and energy prices were quite low.

It took almost 30 years before interest in ground source heat pumps (GSHP) started up again; J Sumner was the first person to use the ground as a heat pump in the UK (Sumner 1976). In 1940 he used the ground as a heat source in a single storey house; heat was supplied by copper pipes buried in a concrete floor using a horizontal collector at a depth of about 1m.

The first GSHP in North America was installed in Indianapolis on October 1945 (Crandall 1946); copper pipes were buried at a depth of about 1.5m in the ground with the refrigerant circulating directly through them. Over the next few years different methods of using the ground as a heat source were looked into mainly in the US (Kemler 1947) and a study in 1953 listed 28 different experimental installations.

The oil shock in 1973 sparked interest in commercial use of GSHP systems and by the late seventies there were over 1000 GSHP installed in Sweden (Granryd 1979). The late seventies also saw the introduction of the vertical earth heat exchanger (Rosenblad 1979), which was used in most types of GSHP systems mainly in Sweden, Germany, Switzerland and Austria.

Though there have not been any major technological breakthroughs since the 1980's in the heat pump, there has been improved reliability. The increased popularity of the system has led to research into improving the system such as system integration, reducing costs for the ground heat exchange and control systems. As technology advances, the reliability and efficiency of GSHP systems are improving due to factors such as improved materials and better installation techniques.

1.3 A look at GSHP systems and how they work

A GSHP system is a type of renewable technology that is very efficient at space heating and cooling. They work on the basis that the ground underneath the Earth's surface has a relatively constant temperature throughout the year, at a depth of about 1m the temperature remains approximately 8-12°C, this heat energy is created by solar gains from the sun. A GSHP transfers this heat stored in the ground into a building as well as transferring heat out of a building when required.

GSHP systems are made up of a ground heat exchanger, a water-to-water or water-to-air heat pump, heat distribution system, evaporator, compressor, condenser and an expansion valve.

The ground temperature remains typically 8-12°C throughout the year, GSHP use the earth as a heat source absorbing the heat through a mixture of water and antifreeze, known as the refrigerant, circulating the buried pipes.

Electricity is required to pump the refrigerant through the pipes, but the energy output is a lot more than the energy input.

1.4 Different types of GSHP systems

There are different types of GSHP systems available and they are chosen according to the different situation and factors involved such as ground condition, area of land available, excavation costs, and if there is a well or lake on site. The two basic configurations are ground-coupled (closed loop), which are installed horizontally or vertically, and groundwater (open loop) systems which are placed in wells or lakes.

The ground-coupled system - this is when a closed loop of pipe is installed either horizontally (1 to 2m deep) or vertically (50 to 180m) underneath the ground. A mixture of water and antifreeze circulates these pipes collecting heat from the ground surrounding the pipes or to dissipate heat when required.

The horizontal system requires a large area of land that is relatively free from large rocks because the pipes cover a large area usually between 35 and 50m depending on the energy requirements.

The vertical system can be used in most soil and rock terrain or where there is a limited area of land. A borehole usually between 100 – 150 mm is drilled to the required depth and the pipe inserted. The holes are usually drilled to a maximum depth of 180m because there may be problems with backfilling, static pressure or insertion of the exchanger.

The main factors affecting the amount of energy transfer to the building is the area covered by the pipes and the thermal properties of the ground. Marble is the most conductive material and would provide the most heat energy, while water movement will substantially improve any thermal properties.

In a ground water system pipes are put into a well or lake, usually held down using a weight system, and heat is extracted from the free flowing water that is moving around. These systems are more cost efficient than the closed loop systems because they are better coupled with ground temperatures.

1.5 Worldwide use of GSHP

Currently it is estimated that there are 1.1 million GSHP systems installed worldwide with a capacity of almost 15,300 MWt (Blum 2009). GSHP systems are one of the fastest growing renewable energy technologies in the world with most of this growth happening in the USA and Europe. There has been an annual 10% increase of installed systems in about 30 countries between 1994 and 2004 (Lund 2004). There has been a strong growth in the alpine countries in Europe such as Scandinavia due to cold winters and being located in remote areas away from mains gas. There seems to be the slowest growth in mid-European countries due to relatively cheap mains gas.

Sweden, Norway, France, Germany and Finland are leading the way in the number of heat pumps being sold. In 2007 Sweden sold over 90,000 pumps, Norway and France sold over 60,000 pumps, and Finland sold over 40,000 pumps. In contrast, the UK sold around 3000.

1.6 Advantages and disadvantages of GSHP systems

Advantages

• Reduces primary energy consumption – Only a small amount of electricity is required

• Reduces carbon dioxide (CO_2) emissions – Only a small amount of CO_2 is produced by the electricity required to run the motor

• A reduced heating cost – The energy comes from the ground and does not need to be purchased, especially since the price of oil is constantly increasing

Ground source heat pumps

- No onsite emissions or fuel storage – There is no need for the storage of fuel and there are no emissions that may be associated with other types of renewable energies

- Minimal maintenance – The systems are very reliable and does not require any extra work very often

- The installation is fairy invisible – The system does not provide a lot of visual pollution

- The system will provide the same output even in cold spells

Disadvantages

- Initial cost of heat pumps – The initial cost of the equipment can be expensive especially if boreholes are required

- Various energy output – The energy output depends on the area of pipes and conductivity of the ground

- May require large areas of ground – An extensive heat sink is needed to collect energy

1.7 Reducing carbon emissions

In 1851 the world's population was 1.1 billion, today it has passed 6 billion and is predicted to reach 9 billion by 2050 (Ochsner 2008). Fossil fuels are being burned to produce heating in homes and are one of the largest sources of CO_2 emissions in the world.

The UK government's target is to cut carbon emissions in the UK by 34% by 2020 and by 80% by 2050. Around 15% of the yearly emissions cuts between now and 2020 will be achieved by making homes more efficient and supporting small scale renewable energy.

Ground source heap pump systems can be used to help achieve these targets set by the government as they can reduce more CO_2 emissions compared to alternative fuels. The systems release less CO_2 compared to other energy sources; oil is the most polluting energy source followed by coal and natural gas, with GSHPs producing the least CO_2 emissions.

When deciding what type of GSHP system to install there are two main types to choose from, open and closed loop, and the type of system required is usually chosen according to the area available, heating and cooling patterns of the building, and the ground conditions as different types of soils have different conductivity.

Sweden, Norway and France are leading the way in the number of installed GSHP systems, while the UK is far behind with approximately 400 installed units. Although the number of systems in the UK is low, as more research is carried out and successful uses of GSHP systems in different countries are looked into, the number of installed units should greatly increase.

The government has set ambitious targets to reduce carbon emissions. Introducing GSHP systems into domestic properties can help the government achieve these targets by replacing traditional types of energy supplies with a cleaner renewable energy source.

The aim of this study is to look into the advantages and disadvantages of GSHP systems, to see if they are a viable option to be installed in domestic properties in the UK. This will involve looking into the costs associated with a system including initial cost, running costs and payback period as well as the environmental implications of the systems and if this can help the government achieve their environmental targets.

Since most UK buildings will use the system for heating rather than cooling, as there are not usually large periods of time when the UK is very warm, this study will look into the viability of using GSHP systems for heating in domestic properties within the UK.

Chapter 2

Literature Review

2.1 Market development

Ground source heat pumps have existed for a long time but it is only recently that the systems have started to be installed in large numbers. It is predicted that the total thermal installed capacity of GSHP systems in twelve of the most important world markets is currently 15,300 MWt (PBlum 2009). The USA, Sweden and Denmark lead the way with a combined total of 77% of the total GSHP systems thermal capacity. Contrarily, the number of GSHP systems purchased in the UK is very low compared to other countries.

This slow uptake of GSHP systems in the UK is due to initial costs and lack of guidance/standards (H. Singh, 2009). The installation costs of a new GSHP system are typically more expensive than other types of heating systems e.g. condensing boiler with radiators. This higher price discourages people from looking into purchasing the system especially as it is a relatively unknown technology. The GSHP industry is relatively new compared with traditional heating systems so there are not any established standards for designing and installing the systems. This means it is up to the people installing the system to have their own criteria and this makes it difficult for customers to know if they are getting an efficient design. While these two factors contributed to the slow uptake of GSHP systems in the UK, the cooling function of the system played an important part as it is unlikely to be needed in the colder climate of the UK (Nimmo 2007). The systems are more economically feasible in places like the USA because the heat pumps are used for cooling in the summer as well as heating in the winter compared to just being used for heating in the UK.

As more research is carried out into GSHP systems the number of installed systems will increase (H. Singh, 2009). It is predicted that there will be a large increase in the number of systems being used in the UK due to a number of factors:

Ground source heat pumps

- New materials are being used which increase the efficiency of the pumps and provide a longer working life

- As oil prices increase, the payback time of systems are being reduced and can save a lot of money in the long run

- GSHP systems can help meet government targets to reduce CO_2 levels

- The technology is becoming more popular and companies are starting to provide complete installation services

The UK has been slow to introduce GSHP as a renewable energy source especially in the domestic sector. This trend is slowly being corrected as more systems are being introduced mainly due to research and development in the area as well as seeing the success of the systems in other countries such as the USA.

2.2 Environmental targets

There are many different environmental policies in place around the world to help reduce carbon footprints and respond to climate change. Table 1 summarises the main legislations in place to help reduce CO_2 levels around the world.

Table 1 – Energy and CO_2 policies

Policy	Target
Kyoto Protocol – This is an agreement between worldwide countries to reduce global warming by reducing greenhouse gas emissions	The UK was set a target to reduce CO_2 levels by 12.5% from the 1990's level by 2012
Climate Change Act 2008 – This Act builds upon policies and targets set in the Kyoto agreement and is aimed at having an impact further down the line	The government has set out targets to reduce UK carbon emissions by 80% from the 1990's level by 2050

Energy Act 2007 – This Act sets out national and domestic strategies for the UK's government towards energy, particularly tackling carbon emissions and providing affordable energy	Maintaining the reliability of energy suppliers, ensuring all houses are adequately and affordably heated, cutting carbon emissions by 60% before 2050
Climate change levy 2001 – This was introduced by the UK's government and involves charging tax on different energy uses with varying rates depending on the source of energy.	To encourage businesses and public sectors to consider improving their energy efficiency and ratings

How big a role can ground source heat pumps play in reducing CO_2 emissions? Assuming an average CO_2 emission factor for electricity of 0.414 kg/kWh, the use of a GSHP with a seasonal efficiency of 350% would result in the emission of 0.12 kg CO_2 for every kWh of useful heat provided.

By comparison, a condensing gas boiler (assuming a CO_2 emission factor for gas of 0.194 kg/kWh) operating at a seasonal efficiency of 85% would result in 0.23 $kgCO_2$ for every kWh of useful heat supplied i.e. the CO_2 emissions would be almost double those from the GSHP (EEBPH)

Approximately 526.5 TWh of energy was used in the domestic sector in 2006, and 57.7% of this energy was used for space heating. If a ground source heat pump was used to provide the space heating in the domestic sector with 350% efficiency, then this would provide 236.3TWh of renewable energy and greatly reduce carbon emissions.

There are a lot of environmental targets being set by different groups and governments, this shows this is an important issue to many different groups of people. GSHP systems can be used to help meet these targets as they are designed to reduce the primary energy consumption required for heating. One view is that heat pumps can be used in the most important energy sectors as a substitution for fossil fuels (Ochsner 2008). Although all of the literature reviewed state that GSHP systems would reduce energy consumption it is unlikely that they will completely cut out the use of fossil fuels. The use of

GSHP systems, by reducing primary energy consumption, has the potential to reduce the quantity of CO_2 produced by the combustion of fossil fuels (Gilli et al, 1992). This reduction in CO_2 can help reach targets set and reduce the effect of global warming.

2.3 Incentives to prevent energy waste/utilise GSHP system

Global warming is a "hot topic" at the moment; it seems to be a very important issue that every country is taking an interest in. It is one of the most important topics being discussed at the G20 meetings and the extent of environmental protesting shows the passion and the importance of environmental issues to all. There are certain legislations that have been brought in to help promote renewable energy technology such as GSHP systems.

Value added tax

The value added tax (VAT) has been lowered on certain energy saving materials such as insulation and heating controls if they are installed by a professional installer, from 17.5% to 5%. The purchase of insulation increases the effectiveness of a GSHP system, making GSHP installation a more viable option.

Planning Permission

From April 2008 changes were made to policies and requirements for planning permission regarding the installation of domestic micro generation technologies. The new policy enables many micro generation tech-nologies, including GSHP, not to require planning permission unless it is in a listed building (Worcester-Bosch). These relaxed planning regulations for GSHP encourage people to purchase the systems as there are fewer barriers to planning.

Building Regulations

Part L of the building regulations was revised in 2006 and has an impact upon the use of GSHP systems. The change means that "targets are set, but the building designers have an element of flexibility in how they achieve the target emissions rate" (Carbon Trust).

Ground source heat pumps

Code for sustainable homes

The code for sustainable homes is an environmental impact rating system measuring the sustainable performance of the house. The system award homes between one and six stars in various categories of sustainable design, with a minimum standard that needs to be met with regards to energy and water use.

This code enables potential homebuyers to see how much of an impact the house has on the environment and the potential running costs. A GSHP system will decrease the environmental impact of the house and improve the sustainability rating.

Grants and financial help

The low carbon programme was introduced in April 2006 for home owners interested in generating their own heat or electricity. Home owners can apply for up to £2500 per property towards the cost of micro generation technologies including GSHP.

Since the Low Carbon Buildings Programme launched in April 2006, approximately £25 million in grants has been claimed, £7.5 million of which has helped 4,600 households generate their own clean and green energy and £18 million helping a total of 739 projects on school, community, housing association and business buildings (BERR).

The carbon trust offer interest free credit towards the installation of energy saving equipment such as GSHP systems to encourage people to purchase a system. The Carbon Trust's Energy-Efficiency Loans are unsecured and interest free, with no arrangement fees and a straightforward application process. Loans of between £5000 and £200,000 are available (Carbon Trust).

There are other types of financial help available e.g. from energy suppliers which are in place to encourage home owners to switch to a more renewable type of energy and reduce carbon emissions. This financial help is aimed to help the government reach their environmental targets and to help reduce the effect of global warming.

2.4 Cost of GSHP systems

The cost of a GSHP system is one of the most important factors when deciding whether to install a GSHP or not.

This is an important factor because it allows the home-owner to work out whether the system is financially viable, how long the payback period is and how much money can be saved so it can be compared to the cost of other systems.

Many organisations have recognised GSHP systems as an attractive way to reduce energy use, lowering utility costs and reducing pollution including: National Rural Electric Cooperative Association, Department of Energy and Electric Power Research Institute. These organisations have stated that the high cost of the system is the main reason for a lack of market penetration (Kavanaugh 1995).

The capital costs of a GSHP system are made up of the equipment costs for the heat pump unit, the ground coil, the distribution system, the drilling or trenching costs and installation costs. The ground coil and its installation typically range between 30% and 50% of the total capital cost. The drilling/trenching costs are usually higher than the material costs for the piping depending on the ground conditions, making it important to maximise the heat extraction per unit length of bore-hole/trench (BSRIA).

The cost of a heat pump varies but the average cost in the UK is about £3000, for a 6KW pump (Banks 2008). Another point made by Banks (2008) is that while a smaller 6KW heat pump costs £3000 (£500 per KW), a 20 KW unit costs between £5000-£6000 (£250-£300 per KW). Thus the larger the project, the proportionately lower the capital cost of the GSHP system becomes.

The cost of a GSHP ranges from about £800-£1,400 per kW (EST). This puts the cost of a GSHP system at a much higher price than noted in Banks (2008) so it is difficult to know how much a system will cost.

The operating costs depend on the efficiency of the heat pump and energy required for the building. Most of the energy is taken from the environment, without cost, so the operating costs come from the energy required to run the motor. If the energy used to run the motor is free e.g. solar energy then the operating costs would be zero because free renewable energy would be used to run the motor (Ochsner 2008).

The cost of external ground work, trenches and boreholes, depends on the ground condition and machinery/labour costs. Typically, a 50m-long trench to a depth of around 1.5 m is needed. Otherwise, a number of boreholes up to 100m deep will need to be drilled (Ochsner 2008).

When it comes to the breakdown in costs for a horizontal and vertical GSHP system, the labour costs and heat pump units are the most expensive items, so if you are able to install the system yourself you can save on a lot of the cost.

2.5 Cost feasibility

Since a GSHP is not cheap the user will want to know how economically feasible a system is. There are two main factors that the client should consider; the cost benefit and the payback period (Messenger 2009).

Cost Benefit

Cost Benefit = Cost of existing system – Cost of alternative system

Payback Period

$$\text{Payback Period} = \frac{\text{Initial capital cost of heating system}}{\text{Annual cost savings when compared to alternative system}}$$

The cost benefit allows the client to compare the potential savings that can be achieved by replacing an existing system. The payback period shows the client how long it will take to repay the initial costs of the GSHP system, using the savings made from the reduced running costs.

2.6 GSHP system applications

GSHP systems make use of renewable energy stored in the ground and can be used for space heating, water heating and space cooling. In North America a market developed from the need for space cooling as well as heating whereas in central and northern Europe there is no demand for cooling and heat pumps are typically for heating only (BERR). Borehole heat exchangers allow geothermal heat pumps to offer both heating and cooling at virtually any location, with great flexibility to meet any demands (Hepbasli and Ozgener 2007).

Space heating

Heat pumps can remove heat from the earth and transfer it to a building to be used for space heating.

The refrigerant circulates through the underground piping system, picks up heat from the ground and takes it to the heat pump.

It then passes through the refrigerant-filled primary heat exchanger for ground water or antifreeze mixture systems. The heat is transferred to the refrigerant, which boils to become a low-temperature vapour. The reversing valve sends the refrigerant vapour to the compressor. The vapour is then compressed which reduces its volume, causing it to heat up. Finally the reversing valve sends the now-hot gas to the condenser coil, where it gives up its heat (NRC).

Space cooling

Heat pumps can work in reverse to cool a building by transferring the heat out of the building, where the cooler ground absorbs the excess heat. This is a good system to use in buildings that get very warm e.g. North America as the GSHP system can be used for both heating and cooling, therefore eliminating the need for separate heating and air-conditioning systems. An interesting point is that the energy losses of all the components at heating mode are higher than the corresponding values at cooling mode (Esen and Inalli, 2005). This means the GSHP systems are more efficient at cooling than heating.

Water heating

GSHP can be used to provide water heating but is a very inefficient method of doing so. The systems are not very popular for water heating and are usually only used where there is a low demand for hot water. Hot water usually comes out from the tap between 35°C to 45°C.

GSHP system power outputs are inadequate to provide direct heating of mains water so a system is required. The water is heated using a primary coil or jacket wrapped around a storage cylinder. For most domestic heat pumps the maximum water storage temperature is 50°C so an electric immersion heater is required to provide an increase in temperature so the water can be stored at 60°C as required, to reduce the risk of legionella (Hepbasli and Ozgener, 2007).

GSHP systems are used for different functions for different places around the world. Most places use the space heating function of the system but the cooling function of the system is generally only used in hot climates and the water heating function is not used in many places due to low hot water temperatures being achieved.

2.7 Coefficient of performance

The performance of a GSHP needs to be tested to allow calculations to be carried out such as how much energy is being produced, how much energy is being used, and how large a system is required to provide the required energy. The energy efficiency of a GSHP system is measured by its coefficient of performance (COP). The COP measures the level of efficiency or cooling of a system and it is the quantity of energy produced (heat or cold) divided by the quantity of energy inputted e.g. oil, gas, electricity.

The COP of a geothermal pump is generally between 2.5 and 4.0, which makes it far more energy efficient than conventional heating and cooling systems (RC). This means that for every unit of energy used by the GSHP, between 2.5 and 4.0 units of energy are produced in the form of heat and cold. The highest COP for a closed loop system is 5.5 (H. Singh 2009), but this is likely to increase as more research is done into GSHPs.

The efficiency of a GSHP depends on the difference between the temperature of the source and the temperature of the distribution system. The smaller this temperature difference is the higher the COP of the GSHP will be. This means GSHP systems are most efficient with distribution systems with the lowest delivery temperatures. Under floor heating has the lowest delivery temperature so this is the most efficient distribution system to be used with a GSHP system.

2.8 Air source and ground source heat pumps

GSHP systems circulate a mixture of water and antifreeze around a loop of pipe which is buried in the garden. When the liquid travels around the loop it absorbs heat from the ground and uses it to heat radiators, under floor heating systems and hot water. Unlike GSHPs <u>air source</u> heat pumps (ASHPs) convert energy from the air and increase it to a higher temperature, using a process which is very much like a reversal refrigeration effect. Table 2 shows the main advantages and disadvantages associated with each type of pump.

Table 2 – Advantages and disadvantages of heat pumps

	Air source heat pumps	Ground source heat pumps
Advantages	ASHPs require far less space to install and no excavations; this makes ASHPs more suitable for buildings	The temperature throughout the year is nearly constant so the design can be opti-

	that do not contain a lot of outdoor space	mised. Thus it provides the best all year round efficiency
Disadvantages	Extremely low outside temperatures and severe weather conditions can lead to high operational energy consumption. External fans tend to produce noise pollution. Source temperature is lower during the winter when most heat is required and therefore the performance is inferior when it is required the most.	The capital costs are 30-50% more expensive than ASHPs (Lohani 2009)

GSHPs have a source temperature that is higher than air therefore they have higher efficiencies of 300% - 400% compared with efficiencies of 175% - 250% for ASHPs (Privett 2006). Lohani (2009) agreed that a seasonal efficiency of up to 230% using ASHPs can be achieved; stating that in practise an efficiency of up to 220% is likely to be achieved.

Heat pumps provide cheaper fuel costs compared with gas, oil and LPG. Using a heat pump has cheaper running costs and can save money in the long run even if the initial costs of heat pumps are more expensive than other sources e.g. oil/gas boilers. The running costs of a GSHP are slightly cheaper than an ASHP, so this should be taken into account when deciding which heat pump system to install.

There are different levels of carbon emissions for each type of fuel source – gas, oil, LPG, air source, and ground source; these emissions could potentially be significantly reduced by using heat pumps rather than other fuel types. According to ICS (2008) ASHPs and GSHPs produce between 40% and 65% less carbon than the other types of fuel sources. This shows that heat pumps have the ability to greatly reduce carbon emissions by using a cleaner technology which can help the government reach carbon reduction targets. ASHPs produce 2900Kg of carbon dioxide (CO_2) per year compared with 2300Kg of CO_2 per year for GSHPs. While heat pumps produce less CO_2 than oil, gas and LPG, GSHPs produce 21% less CO_2 than ASHPs.

RGB (2009) have slightly different, but reasonably similar estimates for the potential reduction in CO_2 emissions –they estimate that ASHPs and GSHPs produce between 23% and 55% carbon less than the other types of fuel sources, compared with the 40% and 65% of ICS. According to the RGB (2009) estimates the carbon reduction using heat pumps is less than what ICS (2008) estimates. The biggest difference between the figures is the air source comparison to LPG where there is a difference of 22%.

Ground and Air source heat pumps (ASHPs) are a similar technology, ASHPs use air to provide heating to a building whereas GSHPs use energy from the ground to provide heating. GSHPs are usually used in domestic properties where there is enough room for the system to be installed whereas ASHPs are usually used in industrial buildings where there is less outdoor space available.

Both types of heat pumps can produce significant CO_2 reduction when heating a building compared with oil, gas and LPG systems, with GSHPs producing 21% less CO_2 than ASHPs. The government has set a target to reduce the amount of carbon emitted in the UK, and replacing traditional sources of heating with heat pumps in domestic properties can reduce carbon emissions, helping the government reach its targets.

The running costs of heat pumps is less than oil, gas and LPG systems which have the potential to save money in the long run taking into account the cost of the systems. Since the cost of gas, oil and LPG is rising almost every year the savings in running costs is likely to increase with heat pumps.

2.9 Energy comparisons

Energy bills of households have almost doubled since 2003 as oil and wholesale gas prices increase; this amounts to an increase of nearly £500 from 2003-2008 for gas and electricity (WB 2008). There has been an increase in the annual cost of bills per household between 2003 and 2008, with an increase of almost 50% for gas and electricity during this time period.

There are significant estimated annual heating and cooling costs for a 3792 square metre house using a vertical bore ground source heat pump system and a gas furnace and electrical air-conditioning system. BE (2008) states that the annual savings of using a geothermal system compared to a natural gas and electrical air conditioning system will increase every year due to the increase in energy prices with an estimated saving of £3,536 in 2013. This is due to the large increase in running costs of gas and electric systems compared to the small running cost increase of a geothermal system. This agrees with WB (2008) that the prices of gas and electricity are most likely to increase every year, thus increasing the running costs of gas and electric systems, while the increase in running costs of a GSHP will be a small fraction of the other fuel systems.

The efficiency and standard assessment procedure (SAP) rating of different primary space heating sources can be compared. The SAP is the method used to assess the energy performance of a building. A GSHP system has a much higher efficiency than an oil/LPG/gas boiler and provides a significantly increased SAP value, which would be enough to upgrade the Eco Homes rating from one category to another, making GSHP a perfect option in domestic housing. The SAP rating of 96% for GSHP systems shows that it can greatly increase the efficiency of the house while also providing carbon reductions since the rating takes into account CO_2 emissions.

The effects on the environment of different fuel types can be measured through calculating the emissions of CO_2 in the generation of heat energy. These measurements show that GSHP systems produce the least CO_2 emissions compared with the other sources of energy, and that electricity produces the most CO_2 emissions, producing almost 202% more CO_2 than GSHP systems. Natural gas is the second least polluting fuel followed by LPG, oil, and finally electricity.

As electricity becomes increasingly generated from renewable sources (wind power, hydro schemes, gas-driven microsystems) the CO_2 emissions will be minimised, further reducing the CO_2 emissions from GSHP systems (BDRB 2007).

Campbell (2009) agrees with BDRB (2007) that GSHP systems produce less CO_2 emissions compared to LPG, oil, electricity and gas. He also distinguishes that GSHPs produce almost 6000 Kg less CO_2 than ASHPs throughout the lifetime of the owner. Throughout the lifetime ownership an LPG system is the most expensive followed by oil systems and gas systems.

Energy prices are rising almost every year increasing the running costs of electric, gas and oil systems. Gas has seen the largest increase in price rising by almost 94% from 2003 to 2008, with electricity rising by almost 74% during the same time period.

Electricity is the most pollutant fuel, producing 139 kg/GJ, followed by oil (78kg/GJ), LPG (61 kg/GJ), and gas (53 kg/GJ). Since the government is trying to reduce CO_2 emissions and it is estimated that the domestic sector is responsible for over 25% of emissions in the UK (AECB 2006) then replacing oil, gas and electric systems with renewable energy systems seems to be a viable option to help meet these targets.

Throughout an ownership lifetime a condensing oil boiler would work out the most expensive followed by a condensing LPG and condensing gas boiler, this takes into account installation costs, distribution system, fuel costs, servicing costs, and replacement costs.

Chapter 3

Method

3.1 Aim of the book

The aim of this book is to look into the different factors of GSHP systems to see if they should be widely used in domestic houses in the UK.

> In order to achieve the research goal it is critical to clarify the key questions for which answers are sought and why they are important. Precise key questions determine the focus and scope of study. It will help researchers direct the literature review, the framework for study, tools and techniques, and the analysis
>
> Potter (2003)

To achieve this goal the research will be broken down into several objectives:

1. Looking into the costs associated with installing a new GSHP system; this includes items such as initial costs and payback period. This will include information on the cost feasibility of a new GSHP system and if it is a better option financially compared to existing heating systems.

2. Looking into the environmental implications of a GSHP system and how this can fit in with the governments environmental targets listed in *Chapter Two.* This will include information on the environmental performance of GSHP systems and their carbon emissions.

3. To find out if GSHP systems can provide affordable space and water heating in homes. This involves looking into how much it costs to provide space and water heating to houses and if this is affordable compared to other means of heating.

4. Finding out the advantages and disadvantages associated with GSHP systems. This will include all the main advantages and disadvantages that may affect the installation in a house.

3.2 Study structure

To present the analysis in a logical and coherent way the book has been set up with the following structure: *Chapter One* provided a brief introduction of GSHP systems. More extensive knowledge of the concept was provided in *Chapter Two* which contained a detailed review of the available literature on the topic of GSHP systems. This chapter details the objectives and scope of the study as well as a description of the research methodology. *Chapter Four* presents the findings of the study, and *Chapter Five* analyses and discusses the findings so that conclusions can be drawn. Finally, *Chapter Six* contains the final summary of the study, any limitations of the study, and details recommended areas for further research.

3.3 Literature review

A literature review was carried out in *Chapter Two* which looked into research and information published by other authors from journals, books, magazines, and online information sources. The objective of the literature review was to offer a broad analysis of GSHP systems looking into areas such as market development, costs, environmental issues and GSHP applications.

3.4 Research method

Business research is often believed to consist of collecting data, constructing questionnaires and analysing data. However it also includes identifying the problem and how to proceed in solving it (Ghauri et al., 1995).

Data sources can be described as the carriers of data (information). There are two types of data sources (Ghauri et al., 1995)

1. Primary data (field) is collected specifically for the research project. This will be in the form observations and interviews

2. Secondary data (desk) is collected by others. These include academic and non academic sources

Secondary data such as books, articles, journals, publications and research papers will be used to collect the necessary information on GSHP systems for this study as it is considered the most appropriate research method. The case studies are based on qualitative research i.e. on the presentation and the analysis of findings found in the literature, this will be used to explore the facts as well as provide a detailed account of the GSHP installations. Case studies are used to demonstrate examples of real life GSHP installations and how they have performed.

The presentation of primary sources e.g. graphs, tables will be used in the discussion section to support and summarise data found in the literature.

The case studies used are based on GSHP installations in domestic properties in the UK; they were chosen as they provide the most relevant information for this study as shown below.

Case Study	Description
Copt Hewick (2007)	This involved the installation of eight ground source heat pumps by Harrogate Borough council. This is believed to be the first local authority owned homes in the UK to have ground source heat pumps installed. The main objectives were to: • Trial the installation of GSHPs as an alternative to electric and solid fuel space heating • Test a financial model which enables whole life costing to be used • Evaluate the affordable warmth potential of retro-spective installations
Penwith Housing Association, Cornwall(2004)	This involved fourteen bungalows each fitted with Powergen 'Heat-Plant' heat pumps connected to vertical ground loops, providing affordable space heating (via

	radiator systems) and hot water with low carbon emissions. The main aim was to demonstrate that ground source heat pumps could provide affordable space and water heating in existing homes in a rural area where there was no availability of mains gas
West Grimstead Wiltshire (1998)	This involved the installation of a horizontal GSHP system in a 288m^2 detached family house in West Grimstead.
Thornbury Bristol	This involved replacing the electric storage heaters with a more efficient heating system which included under floor heating in a 3 bed-roomed detached house located in a remote location
Scotland	This involved replacing an old house with a new three bedroom house and replacing the old electric system with a GSHP system to supply all of the energy needs

3.5 Problems encountered

The main problem with secondary research was finding relevant case studies involving domestic properties in the UK, since the technology is not extensive in the UK and there are not many case studies available. Using secondary research means the data collected in this study will be based purely on existing literature. This is a problem because many of the case studies available are very vague and do not contain a lot of information and data.

Chapter 4

Data Section

This chapter will show the details and information of the individual case studies used to collect data and information. The purpose of this information is to investigate the potential for ground source heat pumps in domestic houses in the UK.

4.1 Case study 1 - Copt Hewick (2007)

Background information

This study involved the installation of eight ground source heat pumps by Harrogate borough council into older resident's bungalows in the village of Copt Hewick in North Yorkshire. Harrogate council entered into a partnership with Ice Energy to supply the heat pump equipment and this project is believed to be the first local authority owned houses in the UK to have GSHPs installed.

The main aims were to:

1) Try the installation of GSHPs as an alternative to electric storage and solid fuel space heating

2) Test the financial model which enables whole life costing to be used

3) Evaluate the affordable warmth potential of the installation

The houses used in this study were one and two bedroom bungalows built by the council in 1970 and were of cavity wall construction and contained PVC double glazed windows. Before the installation of the heat pumps the bungalows were upgraded with cavity wall insulation and loft insulation to increase the efficiency of the heat pumps.

Heating options

The bungalows are not connected to the mains gas network and the option to connect them to the network was too expensive, so using mains gas fired boilers was not a practical option.

Using LPG fuelled systems was rejected due to the affordability for the residents and also because it needs to be bought in bulk quantities rather than being on demand.

Oil fired heating was rejected due to escalating costs of heating oil and also because of the bulk nature of required purchase.

ASHPs were not used because it was believed that they would not provide adequate warmth.

GSHPs were considered appropriate as they retain existing energy supplies to the bungalows as well as providing adequate heating for the occupants.

The GSHPs used were rated at 4KW consumption and used an in line immersion heater to make sure that stored hot water was pasteurised weekly as well as providing backup heating in case there was a problem. The ground loop installation was made up of nine panels buried both horizontally and vertically, 1m underground. Heating radiators were installed in all of the bungalows, oversized by 30% compared to a combustion fired system.

Table 3 – Energy rating for GSHP system

Heating type	Existing SAP rating	New SAP rating	Existing CO_2 emissions (Kg)	New CO_2 emissions (Kg)
Coal fired	48	68	5800	1700
Electric	57	68	3200	1700

Table 3 shows the difference in the standard assessment procedure (SAP) rating and CO_2 emissions of the existing system compared with the new GSHP systems. The GSHPs greatly improved the energy rating of the bungalows, improving the coal fired bungalows SAP rating by 20 and improving the electric bungalows SAP rating by 11. This improved SAP rating means the bungalows are more energy efficient than before and waste less energy and therefore produce less emissions. The new system has also greatly reduced the CO_2 emissions, producing 4100Kg less CO_2 compared with the coal fired bungalows and 1500Kg less compared with the electricity supplied bungalows.

Ground source heat pumps

GSHP provides a CO_2 reduction of 71% compared to coal fire heating and a reduction of 47% compared to electric heating, and greatly reduces the amount of CO_2 released into the atmosphere from the bungalows. Overall these figures greatly exceed the 50% CO_2 reductions the council were hoping for.

Before the GSHP installation occupants were spending on average 12 % of their income on the heating bills. Data was monitored over a period of 21 months and the bungalows have experienced a running cost of between 49p and 67p per day for heating and hot water supply; the difference in running costs is due to the size of the bungalows. This is on average 3.8% of the occupant's income.

The main defect of the new GSHP system was from leakages from the ground loop system caused by ground settlement. Four out of the eight ground loop systems had a failure of connections between panels which was caused by differential settlement of the panels. This defect had to be corrected by replacing the rigid panels with a loop of flexible pipe which allowed a degree of movement.

The installation of the GSHP system in the 8 bungalows greatly reduced the CO_2 emissions of the buildings which lead to an improvement in the SAP ratings. The running costs of the GSHPs were very low allowing residents to receive appropriate levels of heating to meet their requirements without compromising their financial well being. An advantage of GSHP systems over other types of fuel systems is that they do not require a fuel tank on site which means there are no combustion or explosive gases within the building.

4.2 Case study 2 – Penwith housing association (2004)

<u>Background information</u>

This was the first project in the UK to fit GSHPs in a group of social housing and was completed in July 2004. Fourteen bungalows were fitted with heat pumps connected to vertical ground loops with the aim of providing both water and space heating.

The main aim of the project was to show that GSHPs could provide affordable space and water heating in existing homes.

Ground source heat pumps

Heat pumps were used as they are the most economically viable renewable heat production technology with cost comparable to most traditional heating systems. They also produce sufficient heat, there does not need to be any gas connections, and they produce low CO_2 emissions and are highly efficient.

Each GSHP extracts about 6 MWh of thermal energy from the ground each year at an average efficiency of 325% (COP = 3.25).

The total cost of the project was £185,100 but being a pilot project the costs were higher than would be expected today.

The GSHPs have greatly reduced the CO_2 emissions of the bungalows as they are the least polluting energy source. GSHPs produce almost 2000Kg of CO_2 emissions a year compared with the second least polluting fuel, gas, which produces almost 3000Kg of CO_2 emissions a year.

Compared to the GSHP solid fuel produces the most CO_2 emissions – almost 278% more than a GSHP system. While other fuel types also produce greater CO_2 emissions than GSHP – LPG (79% more), Oil (89% more), Gas (53% more) and electricity (145% more).

Most of the CO_2 emissions (36%) from the bungalows derive from space heating. This figure represents an average of all the bungalows, so the percentages will vary with the size of the buildings. Lighting and power accounts for the second largest CO_2 emissions with 26%, followed by cooking with 20% and water heating with 18%.

The GSHPs cost an average of £520 to run in the bungalows; this is lower than the estimated costs of using other fuel types.

An electric system is the most expensive way to supply energy to the bungalows and a GSHP system would be the cheapest (gas, oil, LPG and Solid fuels are all more expensive than GSHP but cheaper than electricity). Also, space heating has the highest running cost for every type of fuel.

The residents were happy with the new installed GSHP system which delivered space and water heating at an affordable cost. The new system also reduced the CO_2 emissions of all the bungalows as well as increasing their SAP rating; one bungalows SAP rating was improved from 27 to 73 after the installation.

Ground source heat pumps

Lessons learned from the project:

- Vertical boreholes are extremely good for homes with small gardens and careful management can lead to the reduction of disruption to an acceptable level

- Costs of drilling boreholes can be greatly reduced if they are installed by an energy company due to working in significant volumes

- Although GSHP systems are usually associated with under floor heating they work very well with radiator systems, which are much easier to install in existing houses

- Insulation levels in houses should be increased to the highest practical level to prevent wasted energy

- GSHP systems have low noise levels, low maintenance costs (no regular servicing requirements) and have a long life expectancy

4.3 Case study 3 – West Grimstead Wiltshire (1998)

<u>Background information</u>

This case study involved the installation of a horizontal GSHP system in a 288m^2 detached family house in West Grimstead during January 1998. The GSHP used had an efficiency of 316% (COP= 3.16) but when used in summer for mostly water heating this lowered to an efficiency of 250%.

Table 4 – Heat pump cost break down

Heat pump only (GBP):	1,108 including controls
Installation (GBP):	30 (done by the owner), cold loop 78 (marginal cost estimate for excavator time)
Capital cost (excluding heat pump) (GBP):	662 excluding the under floor heating system and the hot water cylinder

Ground source heat pumps

Table 4 shows that the cost of the GSHP system excluding the hot water cylinder and under floor system was £1878. The installation was done by the owner of the house so this saved a lot of money. The average annual running cost, taken over 3 years, for space heating and water heating was £419 a year.

The CO_2 emissions of the heat pump and alternatives are as follows:

- Ground-source heat pump: 3 600 kg CO_2/year

- All electric (efficiency 100%): 8 590 kg CO_2/year

- Regular oil-fired boiler, (efficiency 79%): 6 390 kg CO_2/year

- Gas-fired condensing boiler (efficiency 85%): 4 260 kg CO_2/year

Installing the GSHP reduced the yearly CO_2 emissions by almost 58% compared with the previous electrical system. The running costs were also significantly cheaper, reduced from £1100 to £419 a year, a reduction of 62%.

Advantages and Disadvantages

The occupants have been pleased with the comfort levels achieved and found the system quiet and unobtrusive. In the first year of operation a daily average indoor temperature of around 18°C-23°C was maintained during the heating season from November to April.

The lifetime of the heat pump system is expected to be at least 20 years, which is a longer expected lifetime than a gas boiler (10 years), oil boiler (10 years) and electric boilers (12 years).

The system has been reliable and the performance has been good up to date. Over its first year of operation the ground-source heat pump provided 91.7% of the total heating requirement of the building and 55.3% of the water heating requirement, although it was sized to meet only 50% of the design heating capacity. The energy consumption was lower than expected during the monitoring period possibly due to the warmer weather than during an average year.

The installation has been low maintenance and up to date the heat pump has not required any maintenance apart from filter cleaning, which is performed yearly.

The installation cost makes up a large part of the price of a GSHP system, and by installing it themselves they have saved a great deal of money, increasing the financial feasibility by greatly reducing the payback period. The system proved more than capable of providing the water heating requirements by exceeding the expected supply as well as meeting most of the heating requirements. The reduction in CO_2 emissions compared to the previous system was estimated at about 58% which is a huge amount considering this works out to be about 4900 kg CO_2/year.

4.4 Case study 4 – Thornbury Bristol

Background information

Mr Hoskins originally had electric storage heaters installed in his 3 bedroom detached house located in a remote location. He decided to replace the electric storage heaters with a more efficient heating system which included under floor heating.

Mr Hoskins researched all the available options with gas being ruled out first due to the remote location of the property. LPG was too expensive and with the increasing costs of oil he did not feel that an investment in this would provide value for money in the long run. He decided that a GSHP system would be the best option as there was enough land to install the ground loop and the heat pump would be compatible with an under floor heating system. He also felt he would be able to receive the financial payback from the system and therefore make the investment worthwhile.

The new GSHP system produces almost 26,000KW per year and costs £7435 in total. This system saves an estimated £700 per year in running costs so has a payback period of just less than 11 years; this means the system would have paid for itself in 11 years. The system also saves an estimated 2,770Kg of CO_2 every year.

The new system provides a good level of heating especially compared with the electric storage heaters and provides plenty of hot water for the property. With the electric storage heaters Mr Hoskins spent approximately £80 a month on his electricity bills, but the first winter that the heat pump was installed these bills dropped to £65 a month.

Other advantages of installing the GSHP system include:

- Heat pumps can add value to the house
- Heat pumps improve the environmental credentials of the property
- Heat pumps are easy to install compared to traditional heating systems (no flues, gas or oil piping. No need for storage tanks, no gas mains connection or additional ventilation).
- Heat pumps provide a cost effective heating solution
- Heat pumps are a tried and tested Scandinavian technology

Mr Hoskins was looking not only for a more efficient system but also one that would be financially feasible. The new GSHP system provided an estimated saving of 2,770Kg of CO_2 per year and is £700 cheaper per year to run, making the payback period just 11 years.

4.5 Case study 5 – Scotland

<u>Background information</u>

Dr Lawn decided to replace an old house with a new three bedroom house and he wanted to make its energy supply as sustainable as possible. Having seen heat pumps working in America he was very impressed with what they could do so when he was building the new house he decided to replace the old electric system with a GSHP system to supply his energy needs.

Dr Lawns GSHP installer and architect worked closely together to ensure that the new house was designed to maximise the efficiency of the heat pump. As heat pumps run at a lower temperature than normal central heating systems, he was advised that the foundations should be made of concrete, and that he would need under floor heating combined with the right insulation in the walls, floors and loft.

<u>Summary</u>

- The heat pump had a heating output of 10KW and had a COP of 4
- The total cost of the system was £13,942

- The energy saving trust provided a grant of £3,823
- The heat pump met 100% of heating requirements
- The estimated fuel savings were approximately £460 per year
- The estimated carbon savings were approximately 7,000kg CO_2 per year

Advantages and Disadvantages

The installation of the system took longer than expected because large rocks were discovered when the trenches were being dug. This increased the cost of the installation by several thousands of pounds making it more expensive than a normal system of this size.

Dr Lawn is happy with the new system as it provides all his energy needs while using considerably less energy than the previous electrical system in the house. He believes even though the capital costs were high the benefits make it a worthwhile and sensible investment. The payback time for Dr Lawn's investment is 22 years but this is longer than would be expected with typical

GSHP system due to extra costs incurred when digging the trenches. The installation was unobtrusive and the system has a very low visual impact as most of the infrastructure is hidden beneath the ground.

Chapter 5

Discussion

5.1 Introduction

The purpose of this study was to see whether GSHP systems could fruitfully be used in domestic properties in the UK. Case studies of GSHP installation in the UK have been used to investigate whether the installation in domestic properties is a viable option. The different areas looked into were costs, environmental impact, affordable heating, and the advantages and disadvantages of a GSHP system.

5.2 Cost

The cost of a GSHP system is a very important factor when deciding whether to install a system or not as it allows the homeowner to calculate whether it is financially viable or not. The price determines how much money is needed upfront to pay for the system as well as the payback period, which will determine how long it will take for the system to pay for itself in savings. The UK currently does not have a lot of installed systems and the main reason for a lack of market penetration is the cost.

The cost of a GSHP system is made up of the heat pump, ground coil, distribution system, drilling/trenching costs, and installation costs.

The ground coil and its installation make up the largest cost, ranging typically between 30% and 50% of the total capital cost. The drilling/trenching costs are usually higher than the material costs for the piping (this is dependent on the ground conditions) making it important to maximise the heat extraction per unit length of borehole/trench.

The bungalows in case study two had a 6KW GSHP installed for £13,221 but this was a pilot scheme and this cost is more than what would be expected with an installation today, this puts the cost of the system at £2,203 per KW.

Case study four was for a three bedroom building and cost £7435 for a 12 KW system, this places the cost at £620 per KW. The payback period for this system was 11 years.

Case study five also involved a three bedroom building installation; the installation cost was £13,942 for a 10KW system which is £1394 per KW. This installation cost more than would have been expected due to unforeseen costs of large rocks in the soil. Due to the extra costs the payback period was 22 years.

Case study three involved the installation of the GSHP system by the owner; this means the system only cost £1878 excluding a distribution system. Due to the low costs the payback period of the system was only 3 years.

The cost of a GSHP system varies greatly from project to project; from the results above the cost of a system varied from £620 - £2203 per KW. This range of cost per KW is a lot greater than the £800-£1,400 per kW estimate by EST and £500 per KW estimate by Banks 2008. The data does agree though with Banks (2008) claim that the larger the project, the proportionately lower the capital cost of the GSHP system becomes.

The installation costs make up the largest part of the total costs – approximately 30%. Case study three shows that the cost of a GSHP system can be greatly reduced if the installation costs are reduced. The owner installed the system himself and the installation only cost £108, greatly reducing the cost of the system to £1,878 excluding the distribution system. This means if developers building new houses have their own team of GSHP installers then the cost of installation will be greatly reduced therefore reducing the total cost of the GSHP system.

The heat pump unit is the second largest cost of the total costs. If developers are working on larger projects which require a number of heat pump units then they are likely to get a discount on any bulk purchase, therefore reducing the costs of a GSHP system in a new house.

5.3 Payback period

Another important issue when deciding the financial viability of a GSHP system is the payback period, this depends on the cost of the system and whether the original heating system was electric, gas or oil.

The payback period of the GSHP systems in the case studies varied a lot:

Case study three had a payback period of only 3 years but this was because the installation was done by the owner and no distribution system was purchased.

Case study four had a payback period of 11 years which seems a reasonable return; this heat pump supplied the most energy (12KW) and has the lowest payback period.

Case study five had a payback of 22 years, but this is due to extra unforeseen costs of the system.

If developers can reduce the total costs of putting in GSHP systems in new build houses by getting the parts of the system cheaper, and reducing installation costs due to economies of scale, then the payback period would be more economical.

5.4 Environmental impact

GSHPs can play an important role in the environmental impact of the domestic sector as they are designed to reduce the primary energy consumption required for heating.

Case study one showed that the coil fired bunga-lows reduced their CO_2 emissions by 4,100Kg (71%) when a GSHP system was installed. The bungalows that were heated using electricity reduced their CO_2 emissions by 1,500Kg (47%) when the GSHP system was installed.

In case study two the GSHP produced 2,000Kg of CO_2 per year. This is a large decrease in emissions compared to the other fuel sources: Gas would produce a 53% increase, LPG would produce a 79% increase, Oil would produce an 89% increase, Electricity would produce a 145% increase, and Solid would produce a 278% increase.

Case study three involved replacing an electric system (8,590Kg) with a GSHP system (3600Kg) this produced a 58% decrease in CO_2 emissions.

Case study four saves an estimated 2,770Kg of CO_2 every year due to the GSHP installation.

Case study five saves 7,000Kg of CO_2 per year due to the GSHP installation.

The installation of a GSHP system in all of the case studies greatly reduced the CO_2 emissions of the buildings, ranging from a saving of 1,500Kg – 7,000Kg per year. This reduction is due to GSHPs being a clean source of energy that only produces CO_2 emissions for the electricity required to run the motor. The most polluting fuel source is solid followed by electricity, oil, LPG, and gas. If all houses within the UK had a GSHP system installed instead of one of the alternatives then much of the highly polluting fuels contributing to the energy consumption of the domestic sector could be replaced with a cleaner energy.

It is estimated that approximately 54% of a domestic properties CO_2 emissions come from space and water heating. With the government target of three million new houses to be built by 2012, by replacing 54% of the houses energy use with a less polluting energy source it will greatly reduce CO_2 emissions, especially considering an estimated 536.5 TWh of energy is consumed in the domestic sector per year.

5.5 Operating costs

The operating costs of a system to provide space and water heating is another important factor in deciding whether ground heat source pumps should be used domestically because it determines how much the bills will cost. These operating costs depend on the efficiency of the heat pump and energy required for the building. With GSHP systems most of the energy is taken from the environment, without cost, so the operating costs come from the energy required to run the motor.

Case study one aimed to evaluate the affordable warmth potential of a GSHP system to eliminate fuel poverty, this is when more than 10% of income is spent on heating bills. Before the installation an average of 12% of the occupant's income was spent on the heating but after the GSHP installation this value dropped significantly to 3.8%.

Case study two had operating costs of £520 per year compared to an electric system that would have cost £1,170, a gas system that would have cost £590, an oil system that would have cost £610, an LPG system that would have cost £865, and a solid system that would have cost £990.

Case study three had an electric system that was costing £1,100 to run per year but the operating cost dropped by 62% to £419 with the installation of a GSHP.

Case study four saved £700 per year by replacing the electric heating system with a GSHP system.

Case study five saved £460 per year by replacing the electric heating system with a GSHP system.

In all of the case studies the occupants reduced their heating costs due to the installation of a GSHP system, providing more affordable space and water heating. The savings ranged from £419 - £700 per year greatly reducing the heating costs. GSHPs are the cheapest to run out of the traditional heating systems and the greatest savings are from the comparison to electric systems followed by solid, LPG, oil and gas systems. Reducing the operating costs of the heating system can help protect vulnerable people from fuel poverty as well as saving occupants money every year.

5.6 Advantages and disadvantages

A number of advantages and disadvantages were learnt during the installation and use of the GSHP systems in the case studies. These are important to look at so lessons can be learnt and problems avoided in future installations.

Advantages

When building new houses, especially in rural locations, the buildings may not be close to the main gas network. GSHP systems do not need to be connected to the mains network to provide heating whereas installing other heating systems may require this gas connection, which can be very expensive to connect to.

Another advantage that GSHPs have over other heating systems such as LPG and oil is that the fuel is not required to be bought in bulk quantities, which requires storage space on site.

The heat pump installations are free of pollution, generally have low noise levels, low maintenance costs (no regular servicing requirements) and have a long life expectancy. There are also no local emissions, although there will be carbon dioxide emissions associated with their pump, but these are much less than other forms of heating, as mentioned above.

The heat pump installations are unobtrusive and have very low visual impact as most of the infrastructure can be hidden beneath the ground.

GSHPs improve the SAP ratings for assessing energy performance of dwellings as well as providing an improved ecohomes code for sustainable homes ratings for new builds. In case study one the coal fired bunga-lows increased their SAP rating from 48 to 68 and the electricity supplied bungalows increased their SAP rating from 57 to 68. In case study two one of the buildings managed to dramatically improve its SAP rating from 27 to 73. This improvement in rating will not only make the house more efficient but increases the saleability.

Although the cost of drilling boreholes can be expensive the cost can be greatly reduced if they are installed in significant volumes. Developers who build many new houses can greatly reduce the installation costs by using their own team and equipment.

The lifetime of the heat pump system is expected to be at least 20 years, which is a longer expected lifetime than a gas boiler (10 years), oil boiler (10 years) and electric boilers (12 years).

As well as advantages for the occupants of the buildings there are also advantages for sellers to install GSHP systems in domestic properties:

• Heat pumps can add value to the house, increasing the profit when selling

• Heat pumps improve the environmental credentials of the property improving the saleability, including to vulnerable people who are worried about fuel poverty

- Heat pumps are easy to install compared to tradi-tional heating systems (no flues, gas or oil piping. No need for storage tanks, no gas mains connection or additional ventilation)

Disadvantages

Installations can take longer than expected due to the condition of the soil, which increases the costs. In case study five large rocks were discovered when digging the trenches and this increased the cost of instillation by several thousand of pounds.

The ground loop system can become damaged and leak the refrigerant, although this is very rare. Case study one had a failure of connections between panels which was caused by differential settlement, and had to have the ground loop replaced which was very expensive.

The installation can be intrusive and be noisy caus-ing disruption to any occupants or neighbours. Adequate garden space needs to be available to store the system as well as provide room for any machinery required.

Chapter 6

Conclusions

6.1 Introduction

Most of the data collected from the case studies has been positive and there has been positive feedback from the people who had the system installed. There were a few problems with the system mainly with the pilot schemes, but this was to be expected from a pilot and it allowed lessons to be learned. Considering both the advantages and disadvantages from this study the results suggest that ground source heat pumps can fruitfully be used in domestic properties in the UK.

Currently the number of installed units in the UK is at a very low level compared to many other countries, mainly due to lack of market penetration and initial costs. Now that there is more emphasis on using renewable energy sources, this relatively new addition to the domestic heating market can provide cleaner energy.

The system is not fully independent from external power supplies, power is required to drive the pump, but it does use a previously untapped renewable energy source and therefore greatly reduces demand from mains energy sources. With the government's target of cutting down CO_2 emissions in the UK by 34% by 2020, with 15% of the cuts coming from the domestic sector, this has emphasised the need for more renewable technologies. This will increase the number of companies selling and installing GSHP systems, and promoting this technology to developers and individuals.

6.2 Cost of a GSHP system

The total cost of a GSHP system seems to be a barrier to installations and the cost varies a lot depending on a number of factors such as sizing, ground conditions, and efficiency. The main cost comes from installation but if developers have their own team of installers then this cost can be driven down in new build houses. Although the initial costs can be quite high there are savings to be made in the long run especially by replacing inefficient heating systems. There are grants and financial help available to help pay for the systems and encourage their uptake but these are not very well known and many homeowners are missing out.

6.3 Environmental impact of GSHP systems

It can be seen from the data that installing the system can greatly reduce CO_2 emissions compared with installing other traditional types of heating systems. The data suggests that the heat pumps can reduce the CO_2 emissions of houses by more than the 15% target set by the government. Taking into account the government's target of building three million new houses in the UK by 2012, the heat pumps provide the technology to decrease the release of emissions from the new houses compared to the more polluting energy systems. If the government forced a certain percentage of the new houses to be built with a GSHP system, as well as encourage existing houses to be fitted with the system, the CO_2 emissions in the UK would be greatly decreased.

6.4 Running costs

Fuel prices have been rising over the past decade, in 1999 oil cost $18 a barrel and gas cost $1.27 a barrel today these costs are typically around 400 per cent greater; this trend of increasing prices is predicted to continue into the future. With this increase in fuel prices the cost of running traditional heating systems has been rising over the past decade, making it more expensive to heat houses and forcing more people below the fuel poverty line. GSHPs only use a fraction of the fuel other systems use so the running costs are much cheaper and any price increases are rising at a fraction of the other heating systems. This means GSHP systems in houses will provide a cheaper way of heating the house, providing affordable heating that can help keep people above the fuel poverty line. Also the energy price increase reduces the payback period of the GSHP system making it a more attractive prospect.

6.5 Seller advantages

Installing a GSHP system in a house not only provides advantages for the occupants of the building but also provides benefits to the seller. The system improves the efficiency of the house, increasing the SAP rating and provides a cost effective heating solution. These benefits can add value to the property and make it more attractive to homebuyers, thereby increasing saleability.

6.6 Disadvantages

Sometimes there are unexpected costs that may increase the price of the installation, this is mainly when the ground where the loop system is being installed is found to be made up of a composition (e.g. large rocks) that was not expected and will cost more to dig through. This does not happen very often if a professional installer is hired as soil tests should be carried out before installation so that the right equipment can be used.

Another problem that may occur is that the ground loop can become damaged causing the refrigerant to leak into the soil, this would cause the system to not work. This happens very rarely as the materials used for the ground loop should be strong enough to cope with all ground conditions.

6.7 Governments role in encouraging GSHP installations

The government can help in a number of ways to increase the number of GSHP systems installed in the UK. There are grants available to help towards the costs. With installation costs being one of the most influential factors when deciding whether a GSHP system should be chosen over other heating systems, a number of steps can be taken by the government to help encourage people to install GSHPs.

The government could increase the amount of money available under grant schemes, helping to cover more of the installation costs. This would decrease the payback period and therefore make it a more attractive prospect.

The GSHP market should be made more competitive to drive down costs.

The reduction in cost (VAT) for energy saving materials should continue and possibly be increased, making

GSHP installations a more effective and cost efficient option.

As mentioned above installation costs make up a large part of the total costs, so research into the drilling methods for the ground loop should be carried out with the aim of finding more efficient/cheaper methods to help reduce costs.

The technology is reasonably new and unknown, so the government should help increase the awareness of the systems, both to the public and developers, through promotion.

The government can help play an important role in increasing the use of GSHPs in the domestic sector in the UK. This would help reduce the domestic sectors reliance on primary energy sources and replace it will a more efficient and less polluting energy. This would not only save homeowners money on running costs but also reduce CO_2 emissions in the UK, helping the government to meet their environmental targets.

6.8 Limitations

Secondary data was used to collect the information for the study but there was a limit on the information available. The information needed was for case studies based on domestic properties in the UK, this limited the data available considerably and only five case studies were suitable; more case studies would have been ideal. This lack of information limited the comparable data that was required e.g. there was not much data to compare the cost of a GSHP system.

Another limitation of the study was that the data used was secondary so specific questions could not be answered, only data in the case study could be used. Most of the case studies had a lot of general information about the installations but important specific data was not found in a lot of the studies e.g. the COP of the GSHP systems were only found in a couple of case studies.

Due to GSHP systems being a relatively new technology the long term use of them could not be looked into. The short term use could be analysed but the long term prospects might not be the same and might alter the financial viability of the system.

This study focuses on closed loop GSHPs, as they are the most common installation, but open loop systems work in a slightly different way and their data would have produced different results. This study didn't take them into consideration when discussing the use of geothermal energy in domestic properties in the UK, so the conclusion might have differed if they were considered.

6.9 Further research

This study contains information on installing GSHP systems in domestic properties and how they can help the government reach their CO_2 reduction targets. The industrial sector also contributes a large amount of CO_2 emissions in the UK, so looking into the possible use of GSHPs in industrial buildings would further the research. This would help the government reduce pollution emissions even more in the UK as well as make buildings less dependent on primary energy consumption.

This study focuses on the short term use of GSHP systems as the technology is relatively new, so there is not a lot of data available on their long term uses. This data would provide valuable information on the long term financial viability and the long term maintenance of the systems as these could alter the expected payback period.

6.10 Overall conclusion

It can be seen that the UK's uptake of GSHP installations is far behind that of many other countries. This technology has been slow to enter the UK's market and there has been limited promotion and interest in the product.

This study has shown the many advantages the system can bring to the domestic sector and why they could be fruitfully installed in domestic properties. Not only will it help to reduce the running costs for homeowners but it will also help the saleability of the property for the sellers, as well as helping the government to cut CO_2 emissions. Considering that as more and more research is carried out that the efficiency of the GSHPs will increase, along with the fact that the price of fuel is increasing, the payback period is decreasing all the time. This is starting to make the technology more appealing to homeowners as well as to developers.

Ground source heat pumps

The government can and should help encourage the uptake of this renewable technology by providing awareness and funding to persuade the domestic market to convert to GSHP systems.

It is likely that the number of GSHPs in the UK will greatly increase in the upcoming years, the main driver being the government trying to cut the UK's CO_2 emissions to comply with their international targets.

Bibliography

ACS, 2009. Renewable solutions, [Online]. Available at: http://www.acsrenewables.co.uk/?page_id=282 [Accessed 10 January 2010].

AECB, 2006. Minimising CO_2 emissions from new homes, [Online]. Available at: http://www.aecb.net/PDFs/NewHomesCO2Savings25May06.pdf [Accessed 21December 2009].

BDRB., 2007, Benefits drivers and barriers in residential developments.

BE, 2008, Bomar energy including geothermal, [Online]. Available at: http://www.bomarenergy.ca/geo.html [Accessed 05 December 2009].

Rawlings, R., Ground source heat pumps technology, Building services research and information association.

Campbell., 2009. Argyll geothermal, [Online]. Available at:
http://www.argyllgeothermal.co.uk/geothermal_heating_cost_chart.html
[Accessed 20 December 2009].

CT, 2009. Carbon trust, [Online]. Available at:
www.carbontrust.co.uk
[Accessed 25 Otober 2009].

Crandall, A.C., 1946. House heating with earth heat pump, New York

BERR.,2010. Business enterprise and regulatory reform, [Online]. Available at:
www.berr.gov.uk
[Accessed 03 January 2010].

EST,2005. Energy saving trust, [Online]. Available at:
http://www.energysavingtrust.org.uk/uploads/documents/myhome/Groundsource%20Factsheet%205%20final.pdf
[Accessed 02 October 2009].

GILLI, P. V., 1992. The impact of heat pumps on the greenhouse effect, Sittard.

GLE, 2008. Green living energy, [online] available at:
http://www.greenlivingenergy.co.uk/faq.html
[Accessed 22 October 2009].

Granryd, E., 1979. Ground source heat pump systems
in a northern climate. Venice, paper No. E1-82.

Hepbasli, A. Ozgener,O., 2007. Modeling and performance evaluation of ground
source (geothermal) heat pump systems.

Esen, H. Inalli, M.,2005. Seasonal cooling performance
of a ground-coupled heat pump system in a hot and
arid climate.

ICS, 2008. Reducing heating costs with a heat pump,
[online] available at:
http://www.icsheatpumps.co.uk/residential/reduce_y
our_heating_costs.php
[Accessed 29 October 2009].

Kavanaugh, S., 1995. Cost containment for ground
source heat pumps, Alabama.

Kemler, E. N., 1947. Methods of earth heat recovery
for the heat pump, New York.

Lohani, S.P. Schmidt, D., 2009. Comparison of energy and energy analysis of fossil plant ground and air source heat pump.

Lund, J. W., 2004. Geothermal heat pumps an over-view. Geo-heat center quarterly bulletin, 22/1, Klamath Falls.

Messenger, J., 2009. An understanding of how to further promote the use of ground source heat pump systems in the UK. Bsc. Reading, Reading University.

Nimmo, K. McChesney, I., 2007. Heat pump survey a report prepared for the community. New Zealand.

NRC, 2009. Heating and cooling with a Heat Pump, [online] available at:
http://oee.nrcan.gc.ca/publications/infosource/pub/home/Heating and Cooling with a Heat Pump Section4.cfm
[Accessed 18 December 2009].

Ochsner, K., 2008. Geothermal heat pumps, Earthscan, UK.

Blum, B., 2009. Renewable energy 35, pp.122–127

Privett, B., 2006. Heat pumps what are they and why should we use them.

Potter, S., 2003. Heating phenomenon. Phd. London, Brunel University

RC, 2007. Renovables Combinados, [online] available at: http://www.renocombi.com/geoenergy2.html [Accessed 29 October 2009].

RGB, 2009. Heat source pump systems, [online] available at: http://www.rgb-services.co.uk/hsps.html [Accessed 29 October 2009].

SOLO, 2009. Heating installations, [online] available at: http://www.soloheatinginstallations.co.uk/ground_so urce_heat_pump.htm [Accessed 22 October 2009].

Singh, H. et al., 2009. Factors influencing the uptake of heat pump technology by the UK domestic sector, Renewable energy.

Sumner, J. A., 1976. Domestic heat pumps, Prism press, UK.

WB, 2008. rising energy prices, [online] available at:
http://www.worcester-bosch.co.uk/homeowner/our-company/news/rising-energy-prices
[Accessed 29 October 2009].

Wirth, P.E., 1955. History of the development of heat pumps. Bauzeits 73. pp. 47-51.

Lightning Source UK Ltd.
Milton Keynes UK
UKOW041359210512

193006UK00008B/14/P